YOUR KNOWLEDGE HAS VALUE

AF302330

- We will publish your bachelor's and master's thesis, essays and papers

- Your own eBook and book - sold worldwide in all relevant shops

- Earn money with each sale

Upload your text at www.GRIN.com and publish for free

Imprint:

Copyright © 2016 GRIN Verlag, Open Publishing GmbH
Print and binding: Books on Demand GmbH, Norderstedt Germany
ISBN: 9783668353930

This book at GRIN:

http://www.grin.com/en/e-book/344989/the-ring-pendulum-a-physics-exploration-of-diameter-and-time-period

Sumaanyu Maheshwari

The Ring Pendulum. A Physics Exploration of Diameter and Time Period

GRIN Publishing

Table of Contents

Sumaanyu Maheshwari

<u>Diameter and the Time Period</u>
— A physics exploration on ring pendulum

Image source:
http://cs.brown.edu/people/orgs/artemis/old/2014/StudentProjects/AshleyLeonardo-
3/html.finalproject/index.html

<u>The Research Question</u>
In the course of my day to day life, I have watched many objects and systems in oscillatory motion and have been contemplating about them almost every single day. One day, as I sat on my chair, looking at the ring in my hand oscillate, I wondered why its time period was so fast. I asked myself, "Does it depend on the size of the ring?" Being an avid fan of amusement park rides, I was then compelled to relate it to thrilling rides like Disk'O and Pirate Ship. Even though the shape was not the same, my interest insisted me to make the observations for the same. What I asked myself proved out to be true. I saw that the ring being smaller in size takes lesser time and the amusement park rides being greater in size took longer time. My qualitative observations forced me to find the quantitative results. My research question thus asks *"To what extent does the diameter of the ring pendulum affect the time taken to complete one oscillation at constant linear mass density?"*

<u>History</u>
Time is a physical quantity that can be measured by using a phenomenon that is repeated at constant frequency, such as the sunrise and sunset; and phases of the moon. One can also relate this phenomenon to the oscillation of the pendulum. Throughout history various experiments have been carried out to measure this quantity.

When I looked at the antique pendulum watch at my home I pondered if it worked on the same principle. I made some research about it and I found that the concept of measuring time and such clocks have a very interesting background. Galileo's research on the properties of pendulum in 1602

Sumaanyu Maheshwari

is considered to a milestone in the construction of the pendulum clock. His interest in this was sparked when he looked at the back and forth motion of a lamp in the Pisa cathedral[1].

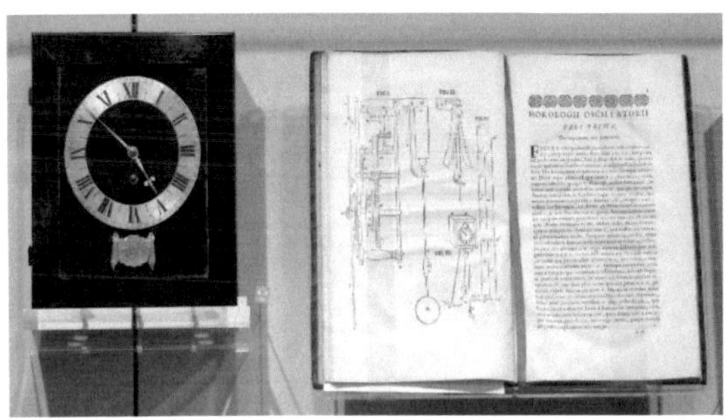

Imagesource:https://www.flickr.com/photos/koopmanrob/3775343287/sizes/o/

Later the pendulum clock was invented by the Dutch scientist Christiaan Huygens in 1656 on the same grounds of Galileo's research.

Hypothesis

From the pre lab activity, I noticed that time period of such a ring pendulum must depend upon its size (diameter). My research question was bolstered by an intriguing question that I had in my mind. I asked myself again "What is the exact kind of relationship between a ring's diameter and its time period of oscillation?" Thus my aim rested on the mathematical grounds to find out the exact relation.

For it I preferred to plot a graph between time period of the ring pendulum against the diameter of the ring with an expectation for the directly proportional relation between them.

If there would not be direct proportionality between them then I preferred the second approach as given below:

I hypothesized an empirical relationship for the time period of this pendulum with the diameter of the pendulum as stated below

$T = k \times D^p$

In the above equation:

T is the time taken to complete one complete oscillation, in seconds;

D is the diameter of the ring, in cm;

P is the constant exponent of the diameter;

K is the proportionality constant.

---

[1] Helden, Al Van. "The Galileo Project." *Http://galileo.rice.edu*. Rice University, 1995. Web. Oct. 2015. <http://galileo.rice.edu/sci/instruments/pendulum.html>

Sumaanyu Maheshwari

I have raised the diameter to an unknown power, to me, as of now because I am unsure of the exact relationship between my *independent variable, the diameter and the dependent variable, the time period*. This value of P can have any value ranging from integers to fractions.

Preliminary Work

Ring pendulum may oscillate in a number of ways. It may oscillate about a point on its circumference in its vertical plane or to and fro perpendicular to its vertical plane about the same point.

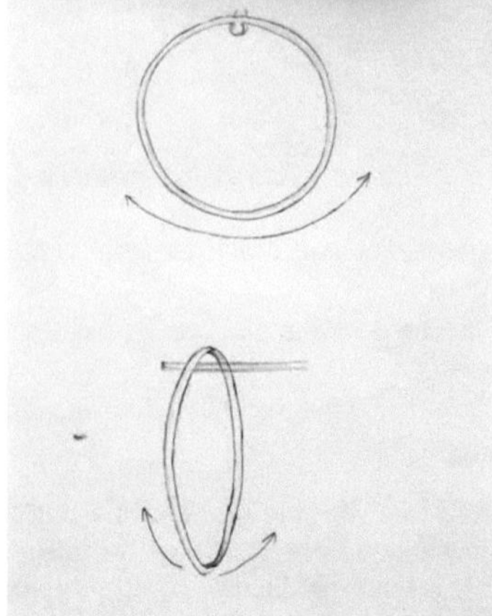

Drawing showing possible modes of oscillations of rings

I decided to investigate about the time period for its oscillations about a point on its circumference in its vertical plane.

1. *Choosing the variables and constants:*

Since in my experiment I was finding out the relationship of the diameter and the time taken one complete oscillation, I took the diameter of the ring as an independent variable. According to my hypothesis the time period must depend on the diameter, although unknown of the exact proportionality, I chose the time period of oscillation as a dependent variable.

I was not sure about the dependency of time period of this ring pendulum on the density of the material of the ring but it might be a hidden factor hence I decided to keep the linear mass density constant by using same material for the rings through-out my experiment.

2. *Choosing the material for the wire of the ring:*

I had a variety of materials to choose ranging from iron to copper for the wire of the ring. as per their availability. Materials like aluminum, brass were also available. I decided to go for Iron. The reason to choose iron was that it is highly attracted to an electromagnet, which may be helped release the ring from rest. I also chose Iron as it was easily available and was cost effective.

I knew forming a perfect ring could be extremely difficult, however I chose to weld the iron wire at both the ends to form ends to form nearly a perfect ring.

3. *Choosing the diameters for the rings:*

I had to choose such diameters for the rings that could easily be measured and through which the experiment could be performed in an effective and efficient manner. For this I conducted a pre-lab with a ring of diameter 2.0 centimeters. I came to a conclusion that this size of the ring, was not practical for my experiment as its time period was too small to observe and get proper values. So I decided to make the starting diameter of the ring as 5.0 centimeters. I took the diameters till 21.0 centimeters with the successive increment of 2.0 centimeters. I measured these diameters through a meter scale.

I had another choice to use Vernier caliper to measure the diameter of the ring but its range was 0 to 15.0 centimeters only and I wanted to collect the data in a range from 5.0 to 21.0 centimeters. Hence it was difficult to control independent variable by using the same Vernier caliper.

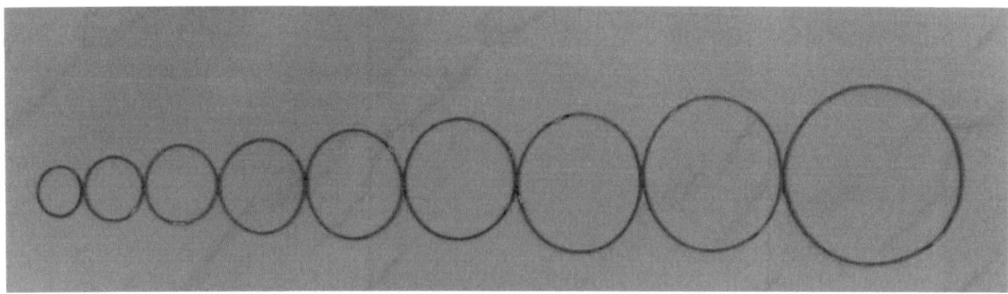

Image showing all the nine rings in ascending order

4. *Choosing method to measure the time period of the oscillation*:

In my experiment, when I released the ring for oscillation by using an electromagnetic assembly, it swung to and fro about its equilibrium position. The time taken for the ring to complete one oscillation is called the period of oscillation. I measured the time period manually by using a digital stopwatch. I started the stopwatch when the ring stabilized on its plane. I did so as to get a perfect plane for oscillation as in the pre-lab I noticed that the ring was not stabilized at this plane and was shifting. The instability was probably because of the nearly imperfect shape of the rings even though I made the rings almost circular. I stopped the stopwatch exactly at the moment when it completed 10 oscillations.

It was a sufficient range for the data collection for time period because in pre lab work I noticed that after it the amplitude of oscillation was small enough and it was quite difficult to collect the data with accuracy.

5. *Controlling the way by which the ring is released:*

I controlled the releasing of the ring by using an electromagnet. It is often observed that one cannot release the ball by hand with the same magnitude of force with controlled velocity for every trial. Thus, by using the electromagnet I released the ring at almost same initial velocity that was 0 m/s. This reduced the chance of human error and curbed this error that subsequently effects the time period.

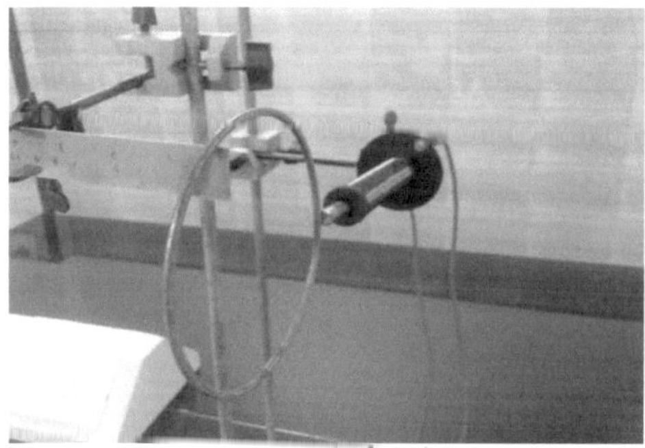

Image showing a ring attached to the electromagnet

6. *Controlling the maximum displacement of the ring:*

In pre lab work I noticed that the ring was slipping on the pivot for large amplitude of oscillation and in that case there was large damping but for small amplitude it was almost constant. So I decided to control the amplitude of oscillation nearly one centimeter. For it I fixed the electromagnetic assembly with a gap of nearly 1.0 cm away from its mean position in horizontal manner along its diameter. Numerical value of the amplitude was taken as an approximation as it was not having significant effect on the time period for small range.

7. Suitable choice for pivot:

I noticed if a thick screw was used as pivot then ring was slipping on it during the oscillation so I made a pivot with knife edge by using a metal scale. It increased the friction on the ring and due to it ring did not slide on the pivot.

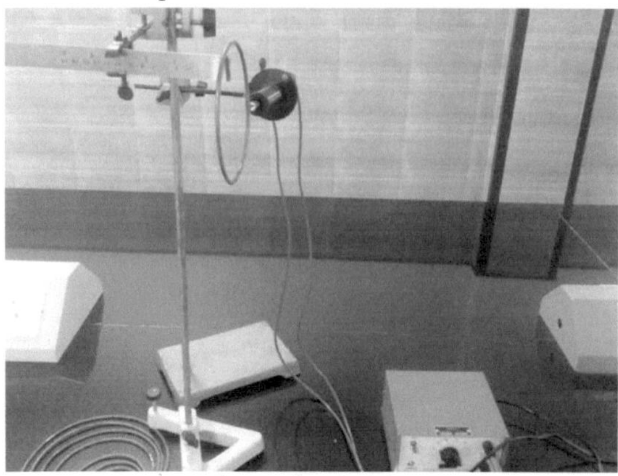

Image showing the setup

Procedure

Firstly, I took all the nine rings and arranged them in ascending order. I then took the clamp stand, and tighten a small metal scale between the clamps of a stand in such a manner that its width side remained vertical and the length side remained horizontal. Width of the metal scale used as pivot was 2.0cm. It could disturb the oscillations of the ring. I took the ring with diameter 5.0 centimetres

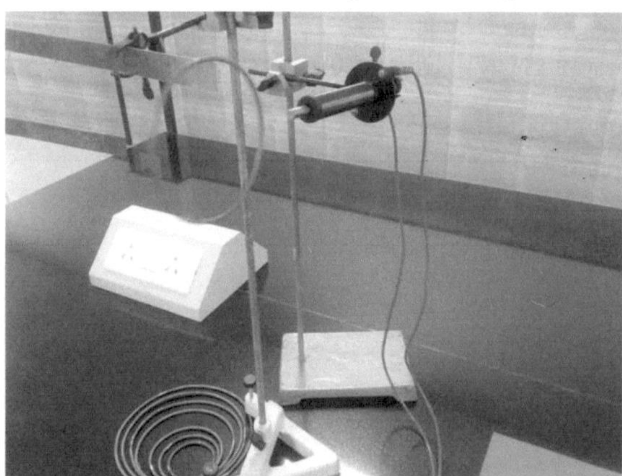

Image showing the ring in motion

Sumaanyu Maheshwari

and placed it on the knife edge of the metal scale. I then turned on the electromagnet and withheld the ring at its maximum amplitude nearly 1.0 cm. I then turned off the electromagnet and instantaneously the ring was released. At the moment the ring stabilized at its plane, I started the digital stopwatch and took the readings till 10 oscillations.

I repeated the experiment for the same ring 4 more times so as to take a total of 5 readings. I then repeated the experiment for ring of 7.0 centimetres, 9.0 centimetres, 11.0 centimetres, 13.0 centimetres, 15.0 centimetres 17.0 centimetres 19.0 centimetres and then finally for the ring of 21.0 centimetres.

After all the readings I made a table of measured diameters and the time taken to complete 10 oscillations in their respective trials. As I got multiple readings of the time period for each diameter, I later took mean of the time taken every diameter to calculate the average time taken for 10 oscillations. I then calculated the time period for one oscillation.

Gathered data

Table 1: Data for diameter and time period

S. No.	Diameter of the ring/ +/-0.1cm	Time for 10 oscillation /s					Average time /s	uncertainty in average time for 10 oscillations $\pm \Delta t$/ s	Time period for one oscillation T/ s	uncertainty in T $\pm \Delta T$ /s
		Trial 1	Trial 2	Trial 3	Trial 4	Trial 5				
1	5.0	4.45	4.51	4.54	4.49	4.57	4.51	0.06	0.451	0.006
2	7.0	5.29	5.46	5.38	5.42	5.33	5.38	0.09	0.538	0.009
3	9.0	6.07	6.08	6.05	6.25	6.12	6.11	0.10	0.611	0.010
4	11.0	6.78	6.59	6.71	6.66	6.83	6.71	0.12	0.671	0.012
5	13.0	7.43	7.37	7.34	7.26	7.32	7.34	0.09	0.734	0.009
6	15.0	7.82	7.95	7.75	7.89	7.86	7.85	0.10	0.785	0.010
7	17.0	8.23	8.38	8.48	8.29	8.42	8.36	0.13	0.836	0.013
8	19.0	8.97	8.79	8.73	8.85	8.69	8.81	0.14	0.881	0.014
9	21.0	9.37	9.17	9.22	9.29	9.12	9.23	0.13	0.923	0.013

Data Processing

I calculated the above data values by the following formulae:

- Average time taken for 10 oscillations $= \dfrac{Sum\ of\ all\ trials\ for\ a\ particular\ diameter}{Total\ number\ of\ trials}$

- Uncertainty in Time taken for 10 oscillations $= \dfrac{(Maximum\ value - Minimum\ value)}{2}$

- Time period for 1 oscillation $= \dfrac{Time\ taken\ for\ 10\ oscillations}{10}$

- Uncertainty in diameter = sum of uncertainties in position of each diametrically opposite ends of the ring on scale

- Uncertainty in time period for 1 oscillation $= \dfrac{uncertainty\ in\ time\ taken\ for\ 10\ oscillations}{10}$

- **Uncertainty in diameter:** Since I will be measuring the diameter of the rings from both right and left side, therefore uncertainty in the diameter of the rings,

$$\frac{least\ count\ of\ the\ meter\ scale}{2} + \frac{least\ count\ of\ the\ meter\ scale}{2}$$

= least count of the meter scale = ±0.1cm

Plotting the graph

Using the data from above, I plotted a graph between the diameter (D) and the time period (T) to confirm about the directly proportionally between the said quantities.

Graph1: Diameter of the rings v/s the time period for 1 oscillation

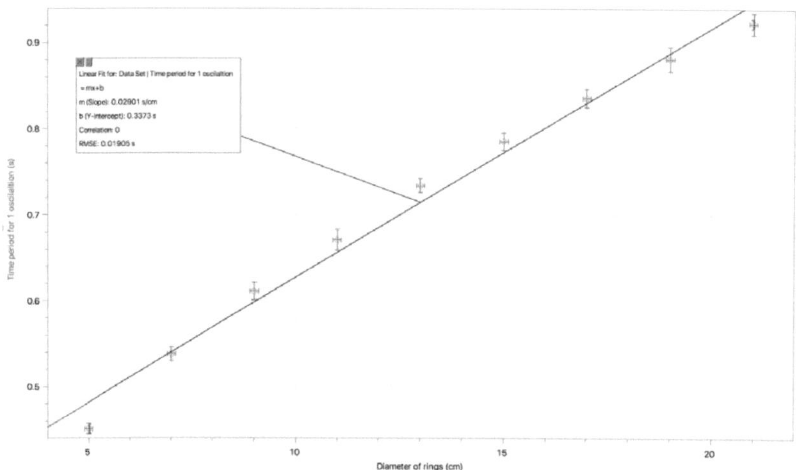

Graph above is a best fit line graph but the data sets are not lying on the same straight line. They are scattered across the line and form a curve line pattern. Since the pattern of the data did not satisfy the condition for straight line, so I decided to use the variable in a manner with mathematical approach for a straight line graph. It was possible if I took log on each side of the empirical relation as I hypothesis so as to get a straight line.

Empirical relation between the diameter of the ring and its time period was $T = k \times D^{p}$

Now, Taking log on each side:

$$Log\left(T\right) = Log\left(k\right) + Log\left(D^{p}\right)$$

$$Log\left(T\right) = Log\left(k\right) + pLog\left(D\right)$$

After comparing this result with standard equation of straight line, I concluded to plot $\log\left(D\right)$ on x-axis and $\log\left(T\right)$ on y-axis.

By doing so I would be able to find an exact value of p, which was resulting the empirical result between the time period and the diameter of the ring.

Processing the data for $\log(T)$ *and* $\log(D)$

Table 2: Data for Log of Diameter and the Time Period for 10 oscillation

S. No.	Diameter (D/+/-0.1cm)	log(D)	$\pm\Delta\log(D)$	T/s	$\Delta T / \pm s$	log(T)	$\pm\Delta\log(T)$
1	5.0	0.699	0.009	0.451	0.006	-0.346	0.006
2	7.0	0.845	0.006	0.538	0.009	-0.270	0.007
3	9.0	0.954	0.005	0.611	0.010	-0.214	0.007
4	11.0	1.04	0.004	0.671	0.012	-0.173	0.008
5	13.0	1.11	0.003	0.734	0.009	-0.134	0.005
6	15.0	1.18	0.003	0.785	0.010	-0.105	0.006
7	17.0	1.23	0.003	0.836	0.013	-0.078	0.007
8	19.0	1.28	0.002	0.881	0.014	-0.055	0.007
9	21.0	1.32	0.002	0.923	0.013	-0.035	0.006

Data Processing

I calculated the above data values by the following formulae:

- Uncertainty in Log(D): $\pm\dfrac{Log(D+\Delta D)-Log(D-\Delta D)}{2}$

- Uncertainty in Log(T): $\pm\dfrac{Log(T+\Delta T)-Log(T-\Delta T)}{2}$

Plotting the graph

Using the data from above, I plotted a graph between Log(D) and Log(T)

Graph2:Log of Diameter of the rings v/s log of the time period for 1 oscillation

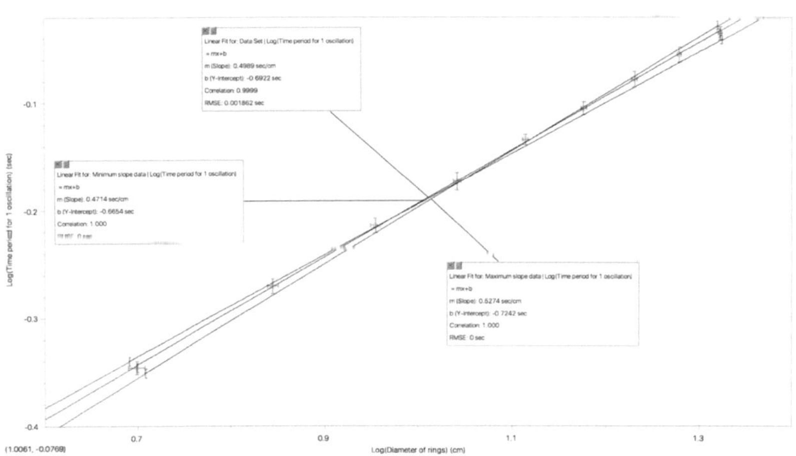

Sumaanyu Maheshwari

From the above graph:

The slope of best fit line is 0.499 s/cm, as per the rules of significant figures

Since all the data sets are lying on the each of the 3 lines, the maximum slope of the graph is 0.5274 s/cm and the minimum slope of the graph is 0.4714 s/cm

The uncertainty in slope is: $\frac{(Maximum\ value - Minimum\ value)}{2} = \frac{0.5274 - 0.4714}{2}$ s/cm= ±0.028 s/cm

Thus the slope of best fit line is 0.499 ± 0.028s/cm

Similarly for y-intercept

The y intercept is -0.692s, as per the significant figures.

The maximum y-intercept is -0.7242 s and the minimum y-intercept is -0.6654 s

The uncertainty in y-intercept is: $\frac{(Maximum\ value - Minimum\ value)}{2} = \frac{-0.7242 - (-0.6654)}{2}$ s $= $ ±0.029 s $\approx 0.03s$

The y-intercept is therefore(-0.692 ±0.03)s

When I account the uncertainty and the value of the slope (n), I can approximate the value of slope to 0.5 s/cm. Thus, analysing the graph and comparing it to the straight line equation I get:

Log(T) = -0.692 + 0.5Log(D)

To find out the value of k:

Log(k) = -0.692

$k = 10^{-0.692}$

$k = 0.204\ sm^{-1/2}$

Creating a formula for the exact relation:

I obtained the equation Log (T) = -0.692 + 0.5Log (D)

Now, Taking anti-log for the equation I get : $T = k\sqrt{D}$

Substituting the value of k,

Therefore the formula for the time period for a particular diameter, in centimetre, is:

$T = 0.204\sqrt{D}$ (Eq.1)

Data Analysis:

If I now compare the equation I created and the standard equation for the time period of the ring pendulum, $(T = 2\pi\sqrt{\frac{2r}{g}})$ [2] I should be able to find the standard value of "g". This approach will enable me to decide to what extent the equation I found made holds true.

So, $k = \frac{2\pi}{\sqrt{g}}$

$\Rightarrow k\sqrt{g} = 2\pi$

$\Rightarrow \sqrt{g} = \frac{2\pi}{k}$

Since I know the value of k from my equation, I can substitute its value in the above equation

---

[2]GERAIS, ARTIGOS. "A Measurement of G with a Ring Pendulum." _A Measurement of G with a Ring Pendulum_. Scielo, 28 Sept. 2011. Web. 21 Oct. 2015.
<http://www.scielo.br/scielo.php?script=sci_arttext&pid=S1806-11172011000300001>.

Sumaanyu Maheshwari

$$\Rightarrow \sqrt{g} = \frac{2\pi}{0.204}$$

Squaring both sides

$$\Rightarrow g = \left(\frac{2\pi}{0.204}\right)^2$$

$$\Rightarrow g = 948.64 \text{ cm/s}^2$$

Conclusion:

The experimental value of "g" I calculated was 948.64 cm/s^2 is very close to the standard value of "g" is 980.67 cm/s^2 [3]. There is a small difference in these values of (980.67 cm/s^2 - 948.64 cm/s^2) 32.03 cm/s^2. Keepig in mind that the standard value depends on the geographical locations, and is an average value therefore this differnce is negligiable. I can rightfully conclude that the equation I made, showing the quantitative dependency of time period of the ring on its diameter, holds true. Thus my experiment is a success.

Also, From graph 2, since the slope of the graph is approximately equal to 0.5, therefore the diameter is raised to the positive power 0.5. Since the power is positive it means that the time period is directly proportional to the diameter. Also from the equation 1 (T=0.204$\sqrt{D}$), it is evident that time period is directly proportional to the root of the diameter.

Evaluation:

Limitations

Although I kept in mind all the significant figures and did all calculations according to the rules of the significant figures and took 5 trials for each diameter to reduce the random uncertainties. I also took uncertainties regarding the diameter of the rings and for the time period, there still could have been some uncertainties.

I believe I was not able to start the digital stopwatch at the exact time the ring got stabilized and stop the watch when it completed its 10 oscillations. This could have caused significant difference in the readings. I, thus, can account this uncertainty to human error. Moreover, I did not take into the drag force, so there could have been uncertainties due to the resistive forces of the air and the friction forces. Also, the shape of the rings were not perfectly circular, this could have manipulated my results.

But overall, I am chuffed with my results and my equation as there is only a little difference in the values of the time period from the standard equation.

Improvements

I realised using the digital stopwatch to measure the time period manually could have given rise to uncertainties, so I should have used the photogate to measure the time period of the oscillation. Thus adjusting the photogate in such a manner so that it could cross the photogate during its each successive motion, I could easily get the time period of oscillation using the smart timer mode. This would have curbed the human error to a large extent. Thus the time at which starting and the stopping of the would have been more accurate. Also for the rings, I should have taken the mean diameter. Since the outer diameter and the inner diameter of the ring had a considerable difference

---

[3] "Acceleration Unit Converter." *Unit Conversion.* Advameg Inc., n.d, Web. Oct. 2015. <www.unit-conversion.info/acceleration.html>.

Sumaanyu Maheshwari

it would have been a logical step to do. By using the Vernier calliper to measure the comparatively smaller rings; and using the metre scale again to find out the diameter of the comparatively larger rings, I could take their mean and find an average diameter for the rings. Thus making the data more accurate.

Although I used the same wire there could have been a minor difference in the linear mass density of the parts of the same wire. So I should also have accounted for these uncertainties in the wire. Although the thickness of the rings was almost the same, I did not check and confirm for the thickness. So I should have checked for the thickness and should have made the rings with the same thickness. I tried to make nearly perfect rings, however I could not get the perfect shape for all of them. In addition, I did not confirm for the centre of mass and the centre of gravity, which could have led to discrepancies. Therefore for more accurate values and to conduct the experiment with utmost care, I should have confirmed for both centre of mass and gravity.

Further scope:

I conducted this experiment in the plane of the ring, in the future I can do it out of the plane of the ring to check if the relation I formed is holds true in such a case too. I shall be doing this experiment keeping in mind all the missteps I did in this experiment and shall be doing it with the enhanced methods discussed in the evaluation section.

Sumaanyu Maheshwari

Bibliography

1. Johnstone, Adrian. "Galileo and the Pendulum Clock." Royalholloway, 8 July 2009. Web. 16 Oct. 2015. <http://www.cs.rhul.ac.uk/~adrian/timekeeping/galileo/>.

2. Helden, Al Van. "The Galileo Project." *Http://galileo.rice.edu*. Rice University, 1995. Web. Oct. 2015. <http://galileo.rice.edu/sci/instruments/pendulum.html>

3. Acceleration Unit Converter." *Unit Conversion*. Advameg Inc., n.d. Web. Oct. 2015. <www.unit-conversion.info/acceleration.html>.

4. GERAIS, ARTIGOS. "A Measurement of G with a Ring Pendulum." *A Measurement of G with a Ring Pendulum*. Scielo, 28 Sept. 2011. Web. 21 Oct. 2015. <http://www.scielo.br/scielo.php?script=sci_arttext&pid=S1806-11172011000300001>.

5. Piggott, R. Keith. "ALEXANDER BRUCE'S ENGLISH AND DUTCH LONGITUDE SEA-CLOCKS REDISCOVERED." *A Royal 'Haagseklok'*(n.d.): n. pag. *Antique-horology*. Keith Piggott, 27 Sept. 2012. Web. 9 Oct. 2015. <http://www.antique-horology.org/piggott/rh/appendix5.pdf>.

6. Sciencemuseum. "Huygens' Clocks." *Sciencemuseum*. The Science Museum, n.d. Web. 12 Oct. 2015. <http://www.sciencemuseum.org.uk/onlinestuff/stories/huygens_clocks.aspx>.

YOUR KNOWLEDGE HAS VALUE

- We will publish your bachelor's and
  master's thesis, essays and papers

- Your own eBook and book -
  sold worldwide in all relevant shops

- Earn money with each sale

Upload your text at www.GRIN.com
and publish for free